How Can You Tell?

Natalie Lunis

Contents

Questions, Anyone? 2
Which One Melts Faster? 4
Which One Stays Warmer? 6
Will It Grow? 8
Will It Stay Green? 10
Which One Goes Farther? 12
Success! 14
    Glossary and Index 16

Questions, Anyone?

Can you use salt to melt ice? Can grass stay green without sunlight? Is a lima bean really a seed? How can you tell? One way to tell if your answer is right is to do an **experiment.**

These scientists are doing an experiment to find out more about soil.

Have fun trying the experiments in this book. They will help you figure out how to set up experiments of your own.

Which One Melts Faster?

Have you ever seen people putting salt on icy sidewalks? Maybe you asked yourself, "Why are they doing that?"

Does salt make ice melt? Tell what you think the answer is. Then you can do an experiment to find out if you are right.

Try This Experiment

You need two plates, two ice cubes, a teaspoon of salt, and a watch or clock.

1. Put each ice cube on a plate.

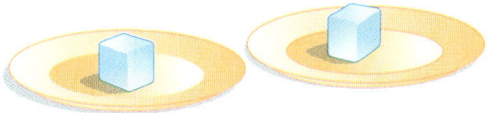

2. Put the salt on one of the ice cubes.

3. Look at the ice cubes every two minutes for ten minutes. What do you **observe**? Do the ice cubes look different from each other?

4. Think about what you observed. Now you can draw a **conclusion.** Start a science notebook. Write or draw about your experiment.

Which One Stays Warmer?

You may have noticed that many containers for hot foods and drinks have covers, or lids. Why is that so?

Do the lids help the foods and drinks stay warm? An experiment can help you find out.

Try This Experiment

You need two plastic jars—one with a lid and one without. You need a pitcher of water that is very warm, but not too hot. You also need two thermometers and a watch or clock.

1. Pour the same amount of warm water into each jar. **Measure** the temperature of the water in each one.

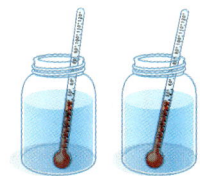

2. Use the lid to cover one jar. Leave the other jar uncovered.

3. Leave the jars in the same place for ten minutes. Then measure the temperature of the water in each jar again. In your science notebook, write or draw what you find out.

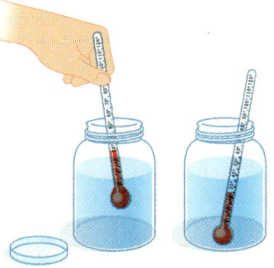

4. Think about what you observed. Draw a conclusion about what it means. **Record** your ideas in your science notebook.

7

Will It Grow?

Did you know that peas, peanuts, and kernels of corn are all seeds? They look different from each other, but they can all grow into plants.

Is a lima bean a seed, too? Here's how to do an experiment to find the answer.

Corn plants

Corn kernels

Try This Experiment

You need a wet paper towel, a clear plastic cup, three dry lima beans, and a spray bottle filled with water.

1. Roll the wet paper towel and place it all around the inside of the cup.

2. Place three lima beans between the paper towel and the side of the cup.

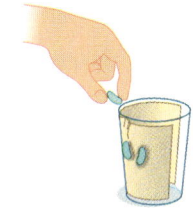

3. For the next week, keep checking to make sure that the paper towel does not dry out. Spray it with water to keep it wet.

4. Observe the lima beans each day. Describe any changes you see. What is your conclusion? Write or draw what you see each day in your science notebook.

Will It Stay Green?

If you look around a grassy park, yard, or field, you can see that grass grows in sunny places.

Does grass need sunlight? Would it stay green and healthy if you blocked out the sunlight for a long time? Do an experiment to find out.

Try This Experiment

You need a grassy outdoor spot, a large piece of cardboard, and a heavy rock.

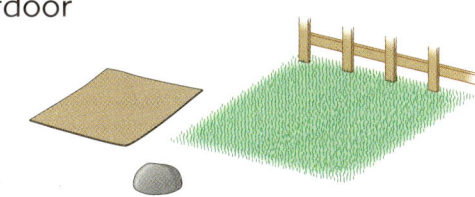

1. Choose a grassy spot that you can visit several times over the next two weeks. Cover some of the grass with the piece of cardboard.

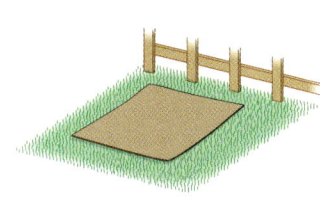

2. Place the rock on top of the cardboard to keep it from blowing away.

3. Every few days, visit your spot and take off the cardboard for a few minutes. Draw or write about any changes you see in the grass.

4. At the end of two weeks, **compare** the color of the covered grass with the color of the grass that was not covered. Draw a conclusion.

Which One Goes Farther?

How is a shopping cart like a baby stroller? They both have wheels. Lots of other things have wheels, too.

Is it easier to move heavy things when they are on wheels? Find out for yourself. Try an experiment.

Try This Experiment

You need a shoe box, rocks, a piece of string about one meter long, a skateboard, and three sticky notes.

1. Put the rocks in the shoe box. Put a sticky note on the floor to mark the starting spot.

2. Give the end of the shoe box a quick shove. Put another sticky note on the floor to show where the box stopped moving.

3. Now tie the shoe box to the skateboard. Put it on the same starting spot. Push the skateboard the same way you pushed the shoe box. Put a sticky note on the spot where the skateboard stopped.

4. Notice the difference between the stopping places. Draw a conclusion about wheels.

Success!

An experiment might not turn out the way you expect. It might answer your question in a surprising way. It might not answer your question at all. That's okay!

You can do the experiment again to make sure. You can change the way you do it. You can think of a new experiment that might answer the question.

As long as you learn something, every science experiment you do is a success.

Glossary

compare (kum-PAIR): examine things to find ways they are alike and different

conclusion (kun-KLOO-zhun): an idea or answer reached by careful thinking about all the facts

experiment (ik-SPAIR-uh-munt): a scientific test to try out an idea or show the effect of something

measure (MEH-zhur): find out the size, amount, or degree of something

observe (ub-ZURV): look closely and carefully

record (rih-KORD): put information down in writing or by drawing pictures

Index

grass, 2, 10–11

ice, 2, 4–5

lids, 6–7

salt, 2, 4–5

seed, 2, 8

sunlight, 2, 10–11

temperature, 7

wheels, 12–13